ANTONY WORRALL THOMPSON

Mediterranean

Photography by SIMON WHEELER

THE MASTER CHEFS
WEIDENFELD & NICOLSON
LONDON

ANTONY WORRALL THOMPSON first made his mark on the London restaurant scene in 1981 when he opened Ménage à Trois in Knightsbridge. Reflecting the vogue of the time for small helpings of beautifully presented food, Ménage à Trois served only starters and puddings; it led to Antony's first book, *The Small and Beautiful Cookbook*.

He is now the creative force behind a number of London's top restaurants, among them dell'Ugo, Bistrot 190, The Atrium, Palio and Drones. He has written several more books, including *Modern Bistrot Cookery*, *30-Minute Menus*, based on his column in *The Sunday Times*, and *Supernosh*, written with Malcolm Gluck, in which light, modern food is matched with easily available wines.

In 1987 he was awarded the 'chef's Oscar' – Meilleur Ouvrier de Grande Bretagne (MOGB) – one of only five chefs to have merited this lifelong title.

Familiar to viewers of BBC–TV's *Ready Steady Cook!*, he has also appeared on *Food & Drink* and *Hot Chefs*, among others.

Photograph by Ben Wright

CONTENTS

In cooking, as in all
arts, simplicity is the
sign of perfection.

CURNONSKY

INTRODUCTION

It has been shown that the Mediterranean diet – with little or no butter and very few rich creamy sauces – can promote good health and extend our life span. I have chosen ten recipes, not just from Italy and the South of France, but also from Spain, Morocco, Greece and the Lebanon, to show how varied Mediterranean cuisine can be.

The recipes take you through the menu for a lunch or dinner. They comprise two dips to go with drinks; the excellent Greek salad, which everyone adores on holiday but nobody can seem to reproduce in restaurants outside Greece; a simple yet delicious fish soup; a pasta dish loved by all; an adaptation of tabbouleh that has proved to be an exceptional winner; a fresh tuna dish with my own version of the Greek skordalia sauce; a classic paella; a non-beef variation on steak tartare; a warming winter stew from Morocco; and a healthy pud using that most Mediterranean of fruits, the fig.

I think you'll find all the dishes approachable and achievable, with none requiring any special chef skills. Each of the dishes tells a story of my Mediterranean experiences and many feature on my restaurant menus, which have given me my reputation for being at the cutting edge of Mediterranean cooking in this country.

TWO DIPS
from the Eastern Mediterranean

BABA GHANOUSH

- 3 LARGE AUBERGINES
- OLIVE OIL
- 4 TABLESPOONS TAHINI (SESAME SEED PASTE)
- 4 TABLESPOONS GREEK YOGURT
- 4 GARLIC CLOVES, CRUSHED WITH A LITTLE SALT
- JUICE OF 1 LEMON
- SALT AND GROUND BLACK PEPPER
- POMEGRANATE SEEDS
- MINT LEAVES

CUCUMBER, GARLIC AND YOGURT

- 450 G/1 LB GREEK YOGURT
- 3 TABLESPOONS CHOPPED FRESH MINT
- 450 G/1 LB CUCUMBER, PEELED, SEEDED AND SLICED
- 2 GARLIC CLOVES, CRUSHED WITH A LITTLE SALT
- GROUND BLACK PEPPER

Serve both dips with a Mediterranean-style flatbread

SERVES 6

For the baba ghanoush: prick the aubergines all over with a fork, then rub with a little olive oil. Wrap each aubergine in foil and grill over a fierce flame, barbecue or chargrill for about 15–20 minutes on each side or until the aubergines feel soft and have reduced considerably in size. Remove the foil, drain off any juices and peel away the charred skin. Leave to cool.

Mash the aubergine pulp with a fork or potato masher; it shouldn't be too smooth. Stir in the tahini, yogurt, garlic and lemon juice. Season to taste. Spoon the mixture into a serving bowl and scatter with the pomegranate seeds and a few mint leaves.

For the cucumber, garlic and yogurt dip: combine all the ingredients, then taste and adjust the seasoning as required.

Leave in a cool place for about 1 hour before serving, to allow the flavours to develop.

GREEK VILLAGE SALAD
(Horiatiki)

4 PLUM TOMATOES, EACH CUT INTO
   6 PIECES
½ CUCUMBER, PEELED AND CUT
   INTO 1 CM/½ INCH SLICES
125 G/4 OZ FETA CHEESE, CUBED
50 G/2 OZ KALAMATA OLIVES,
   STONED
1 SMALL RED ONION, THINLY
   SLICED
2 TEASPOONS FRESH OREGANO
   LEAVES

DRESSING

5 TABLESPOONS EXTRA VIRGIN
   OLIVE OIL
1 TABLESPOON AGED RED WINE
   VINEGAR
1 TEASPOON DRIED GREEK
   OREGANO
1 GARLIC CLOVE, FINELY CHOPPED
4 CAPERS, FINELY CHOPPED
1 ANCHOVY, FINELY CHOPPED
GROUND BLACK PEPPER

SERVES 4

Whisk all the dressing ingredients
together and leave for 30 minutes
for the flavours to meld.

Place all the salad ingredients in
a large serving bowl. Shake the
dressing and pour over the salad.
Toss to combine.

PROVENÇAL FISH SOUP

6 TABLESPOONS GOOD OLIVE OIL

2 LEEKS, FINELY SLICED

4 GARLIC CLOVES, FINELY CHOPPED

4 LARGE RIPE TOMATOES, CHOPPED

1 CARROT, FINELY SLICED

3 STALKS OF FENNEL

1 STICK OF CELERY, SLICED

2 SPRIGS OF THYME

1.5 KG/3 LB FISH (GRONDIN, GURNARD, RASCASSE, JOHN DORY OR OTHER NON-OILY FISH)

GOOD PINCH OF SAFFRON

SALT AND GROUND BLACK PEPPER

2 TABLESPOONS PERNOD (OPTIONAL)

4 TABLESPOONS DOUBLE CREAM (OPTIONAL)

TO SERVE

SLICES OF BAGUETTE, TOASTED

ROUILLE (PAGE 29)

SERVES 6

Heat the olive oil in a large saucepan. Add the leeks, garlic, tomatoes, carrot, fennel, celery and thyme and cook until all the vegetables have softened.

Chop the fish, discarding the eyes and gills, and add to the vegetables. Let them brown very lightly, then add 2.8 litres/5 pints water. Bring to the boil and add the saffron. Reduce the heat and simmer for 30 minutes.

Strain the soup through a fine sieve into a clean saucepan. Pass the vegetable and fish residue through a vegetable mill and add the resulting purée to the soup.

Taste and adjust the seasoning and stir in the Pernod and cream if you wish.

Serve with toasted baguette slices spread with rouille.

PENNE WITH SPINACH,
Gorgonzola and pine nuts

1 TABLESPOON EXTRA VIRGIN
   OLIVE OIL
2 GARLIC CLOVES, CRUSHED WITH A
   LITTLE SALT
½ ONION, FINELY DICED
150 ML/¼ PINT DRY WHITE WINE
450 ML/¾ PINT DOUBLE CREAM
450 G/1 LB PENNE PASTA
85 G/3 OZ BABY SPINACH
125 G/4 OZ GORGONZOLA,
   CRUMBLED
GROUND BLACK PEPPER
3 TABLESPOONS FRESHLY GRATED
   PARMESAN CHEESE
50 G/2 OZ PINE NUTS, TOASTED

SERVES 4

Heat the olive oil in a large saucepan and add the garlic and onion. Cook over a medium heat until soft but not brown.

Add the wine and cook over a high heat until the wine has reduced and become syrupy.

Stir in the cream and continue to cook over a high heat until the cream has reduced by half. Strain the sauce through a fine sieve and return to the saucepan.

Meanwhile, bring a large saucepan of salted water to the boil and cook the pasta until *al dente*.

Fold the spinach and Gorgonzola into the cream sauce and cook until the spinach has wilted and the Gorgonzola has started to melt. Season to taste with black pepper.

When the pasta is cooked, drain and fold into the Gorgonzola cream. Add the Parmesan and stir to combine. Tip the pasta into a large warmed bowl and scatter with the toasted pine nuts.

SALMON TABBOULEH

450 G/1 LB SALMON FILLET, IN ONE
  PIECE, THOROUGHLY SCALED
SALT
175 ML/6 FL OZ EXTRA VIRGIN
  OLIVE OIL
50 G/2 OZ FINE BULGAR WHEAT
  (CRACKED WHEAT)
450 G/1 LB PLUM TOMATOES,
  SEEDED AND DICED
1 BUNCH OF SPRING ONIONS,
  FINELY SLICED
225 G/8 OZ FLAT-LEAF PARSLEY,
  STEMMED AND CHOPPED BY
  HAND
50 G/2 OZ MINT, STEMMED AND
  CHOPPED BY HAND
½ TEASPOON GROUND CINNAMON
½ TEASPOON GROUND ALLSPICE
½ TEASPOON GROUND BLACK
  PEPPER
JUICE OF 2 LIMES
18 SMALL COS LETTUCE LEAVES

SERVES 6

Season the salmon, rub with a little olive oil and grill, with the skin side towards the heat, until the skin is slightly charred and very crispy. Turn the salmon over and grill for a further 2 minutes. Leave to cool.

Remove and dice the skin; set aside. Flake the salmon flesh – not too small – and set aside.

Rinse the bulgar wheat in several changes of water. For the last change of water, leave the wheat to soak for 15 minutes.

Drain the bulgar wheat and combine with the diced tomatoes, spring onions, herbs, spices and lime juice. Gently mix in the salmon. Season to taste.

Arrange the lettuce leaves around a bowl and fill the centre with the tabbouleh. Scatter the crisp diced salmon skin over the top. Use the lettuce leaves to scoop up the tabbouleh.

GRILLED TUNA
with celeriac skordalia, rocket salad

4 TUNA LOIN STEAKS, ABOUT
  175 G/6 OZ EACH
EXTRA VIRGIN OLIVE OIL
SALT AND GROUND BLACK PEPPER
4 LEMON WEDGES

CELERIAC SKORDALIA

325 G/12 OZ CELERIAC, CUBED
125 G/4 OZ POTATO, CUBED
5 GARLIC CLOVES, CRUSHED WITH A
  LITTLE SALT
250 ML/8 FL OZ EXTRA VIRGIN
  OLIVE OIL
85 ML/3 FL OZ WARM MILK

ROCKET SALAD

85 G/3 OZ ROCKET LEAVES,
  WASHED AND DRIED
4 TABLESPOONS EXTRA VIRGIN
  OLIVE OIL

SERVES 4

To make the skordalia, boil the celeriac and potato in salted water until tender, about 20–30 minutes.

Drain and return the vegetables to a pan over a low heat to dry out. Place the warm vegetables in a food processor and blend until smooth. With the machine running, add the garlic and slowly pour in the oil and milk as if you were making a mayonnaise. Season to taste.

To cook the tuna, preheat a grill. Rub the tuna lightly with the olive oil, salt and pepper. Cook for 1–2 minutes on each side, depending on how rare you like the tuna. (Well-done tuna is a waste of good fish, and if this is your preference then you may as well open a can.)

Season the rocket leaves and toss with the olive oil. Arrange the celeriac skordalia on four plates, top with the tuna and serve the rocket on the side, with a wedge of lemon.

PAELLA
with shellfish and chicken

6 TABLESPOONS EXTRA VIRGIN
  OLIVE OIL
12 LARGE UNCOOKED PRAWNS
1 LARGE ONION, FINELY DICED
12 CHICKEN THIGHS OR RABBIT
  PIECES
4 GARLIC CLOVES, CRUSHED WITH A
  LITTLE SALT
2 SPRIGS OF THYME
6 SMALL SQUID, CLEANED AND CUT
  INTO ROUNDS
175 ML/6 FL OZ DRY WHITE WINE
450 G/1 LB CLEANED CLAMS
  (PAGE 28)
450 G/1 LB CLEANED MUSSELS
  (PAGE 28)
600 G/1¼ LB SHORT-GRAIN RICE
PINCH OF SAFFRON, SOAKED IN A
  LITTLE WARM WATER
1 TEASPOON PAPRIKA
4 LARGE TOMATOES, SKINNED AND
  CHOPPED
3 TABLESPOONS CHOPPED FLAT-LEAF
  PARSLEY
SALT AND GROUND BLACK PEPPER

SERVES 6

Heat half the oil in a paella pan or large frying pan and cook the prawns over a high heat for 2 minutes. Remove and set aside.

Add the remaining oil to the pan, with the onion, chicken, garlic, thyme and squid. Fry until all is golden, about 10 minutes.

Meanwhile, bring the wine to the boil in a large saucepan, add the clams and mussels, cover and cook over a high heat until the shellfish have opened. Strain the cooking liquid through a fine sieve and set aside. When cool enough to handle, shell half the mussels and clams; discard the empty shells.

Add the rice, saffron and paprika to the pan with the chicken, stir to combine. Pour in 2½ litres/4½ pints water and cook for 10 minutes over a high heat.

Reduce the heat to medium and cook for a further 10 minutes.

In the last 3 minutes of cooking, stir in the tomatoes, prawns, clams, mussels and parsley. Season to taste and serve immediately, from the paella dish.

LEBANESE STEAK TARTARE

1 SMALL ONION, FINELY DICED
1 GARLIC CLOVE, FINELY DICED
450 G/1 LB LEAN LEG OF LAMB,
   MINCED
1 TABLESPOON GOOD OLIVE OIL
3 TABLESPOONS CHOPPED FRESH
   MINT
½ TEASPOON GROUND CINNAMON
½ TEASPOON GROUND ALLSPICE
½ TEASPOON GROUND BLACK
   PEPPER

TO SERVE
MINT LEAVES
TOASTED ALMONDS
WHOLE TRIMMED SPRING ONIONS
LIME WEDGES

SERVES 4

Put the onion, garlic, lamb, olive oil, mint, cinnamon, allspice and black pepper in a food processor and blend until smooth.

Transfer to a mixing bowl. With wet hands, work the meat until you have a smooth purée, removing any sinew or fat that you may have missed when trimming the meat.

Shape the meat into flat patties and scatter with mint leaves and almonds. Serve with spring onions and lime wedges, and a warm flatbread such as pitta.

MOROCCAN LAMB STEW
with pumpkin and pickled lemon

450 G/1 LB LEAN SHOULDER OF
    LAMB, CUT INTO 2.5 CM/1 INCH
    CUBES
1½ TEASPOONS FRESHLY GROUND
    BLACK PEPPER
3 TABLESPOONS OLIVE OIL
1 ONION, ROUGHLY DICED
4 GARLIC CLOVES, CRUSHED WITH A
    LITTLE SALT
4 TOMATOES, SKINNED AND DICED
1 TABLESPOON HARISSA (PAGE 29)
    OR HOT PEPPER PASTE
400 G/14 OZ CANNED CHICKPEAS,
    DRAINED
325 G/12 OZ TRIMMED AND PEELED
    PUMPKIN, CUT INTO
    2.5 CM/1 INCH CUBES
1 TEASPOON SALT
1 PICKLED LEMON (PAGE 30), FINELY
    DICED
2 TABLESPOONS CHOPPED FRESH
    MINT
1 TABLESPOON CHOPPED FRESH
    CORIANDER

SERVES 4

Coat the lamb in the black pepper. Heat the oil in a large saucepan, add the lamb and cook until it has browned all over.

Add the onion and garlic and cook until the onion has softened and is slightly browned.

Add the tomatoes, harissa and 450 ml/¾ pint water. Bring to the boil, cover and cook over a medium heat for 1½ hours, topping up with water as necessary.

Add the chickpeas and pumpkin and cook for a further 15 minutes or until the pumpkin is tender. Add the salt, lemon, mint and coriander and serve immediately.

POACHED FIGS
with blackberries in red wine

675 G/1½ LB RIPE, UNDAMAGED
  BLACKBERRIES, WASHED
JUICE OF 2 LEMONS
JUICE OF 1 ORANGE
175 G/6 OZ CASTER SUGAR
1 BOTTLE ZINFANDEL RED WINE
16 FIRM FRESH FIGS
85 ML/3 FL OZ CRÈME DE MÛRE
  (BLACKBERRY LIQUEUR) OR
  CRÈME DE CASSIS
  (BLACKCURRANT LIQUEUR)
1 TABLESPOON FINELY CHOPPED
  FRESH MINT

SERVES 4–6

Put the blackberries in a food processor or liquidizer with the lemon and orange juice and blend until smooth.

Strain the resulting purée through a fine sieve into a non-reactive saucepan. Add the sugar and red wine and bring to the boil over a medium heat. Reduce the heat to a simmer and skim off any scum that may have risen to the surface.

When the sugar has dissolved, add the figs, in batches of four, and poach for 4–5 minutes, depending on their ripeness, until just tender. Remove the cooked figs to a glass serving bowl.

When all the figs are cooked, increase the heat and allow the blackberry purée to boil and reduce to about 450 ml/15 fl oz. Leave to cool, then stir in the liqueur and chopped mint and pour over the figs.

This dish is best prepared 24 hours in advance, turning the figs in the liquid from time to time.

THE BASICS

CLEANING CLAMS AND MUSSELS

Rinse in a bowl of cold water with a handful of sea salt. Scrub or scrape the shells to remove any mud or barnacles. Discard any broken clams or mussels and any that are open and remain open when tapped briskly with the back of a knife. Pull out the 'beards' from the sides of the mussels.

CLEANING SQUID

Rinse the squid. Pull the head and tentacles away from the body; the innards and clear, plastic-like 'pen' should come out at the same time – run a finger inside the body to check that it is clean, and rinse briefly.

Cut off the tentacles above the eyes and squeeze out the beak. Rinse briefly.

HARISSA

50 G/2 OZ DRIED RED CHILLIES
2 GARLIC CLOVES, PEELED
SALT
OLIVE OIL

Soak the chillies in hot water for 1 hour.

Drain the chillies and cut into small pieces. Place in a mortar with the garlic and pound to form a coarse purée.

Sprinkle with a little salt, then transfer to a small, sterilized jar and cover with a layer of olive oil. The harissa can be stored in the refrigerator for 2–3 months.

ROUILLE

50 G/2 OZ DAY-OLD WHITE BREAD,
  CRUSTS REMOVED
6 TABLESPOONS MILK
GOOD PINCH OF SAFFRON
1 TABLESPOON HOT WATER
3 RED CHILLIES, FRESH OR DRIED,
  SPLIT, SEEDED AND CHOPPED
3 GARLIC CLOVES, FINELY CHOPPED
1/4 TEASPOON SEA SALT
4 TABLESPOONS EXTRA VIRGIN
  OLIVE OIL

Soak the bread in the milk for 10 minutes, then squeeze dry. Warm the saffron in a metal spoon over a low heat for about 30 seconds, then pound it in a mortar. Pour over the hot water and leave to infuse. Pound the chillies, garlic and salt in a mortar, then add the soaked bread and pound to form a purée. Gradually pour in the olive oil, as if making mayonnaise, pounding until all is blended. Spread the rouille on toasted bread.

PICKLED LEMONS

LEMONS
FLAKED SEA SALT

Without cutting all the way through – stop about 1 cm/ ½ inch from the stem end – quarter the lemons.

Dust the flesh with salt and reshape the lemons.

Place some salt in the bottom of a sterilized, wide-necked jar and pack in the lemons, layering them with salt and pressing hard to release the juice. (The lemons must be tightly packed.) Add more salt to cover the lemons, then cover with a tight-fitting, non-reactive lid and leave for 30 days, turning the jar upside down each day.

Rinse each lemon in cold water before use.

THE MASTER CHEFS

THE MASTER CHEFS

Text © copyright 1996 Antony Worrall Thompson

Antony Worrall Thompson has asserted his right to be
identified as the Author of this Work.

Photographs © copyright 1996 Simon Wheeler

First published in 1996 by
WEIDENFELD & NICOLSON
THE ORION PUBLISHING GROUP
ORION HOUSE
5 UPPER ST MARTIN'S LANE
LONDON WC2H 9EA

British Library Cataloguing-in-Publication data
A catalogue record for this book is available
from the British Library.

ISBN 0 297 83634 X

DESIGNED BY THE SENATE
EDITOR MAGGIE RAMSAY
FOOD STYLIST JOY DAVIES
ASSISTANT KATY HOLDER